In this book we will take you on a journey outside the Earth's atmosphere, to explore the solar system.

In our Solar System alone there are seven other planets many dwarf planets, satellites and moons and the Sun

Our solar system is in a galaxy called the Milky Way. In the Milky Way is it estimated that there are 30 billion solar systems.
Humans think that there are 100 billion galaxies in the universe. This means that the universe is much bigger than we can even imagine

The Sun

The sun is our local star, It appeared about 5 billion years ago. The sun is a big ball of hot gas, a sphere of molten hydrogen. It's a gigantic fireball, the temperature in the center of the sun is about 14 million degrees. When the sun burns, it emits particles called photons, which is called solar radiation. Solar radiation supplies the planets with energy, without it, the Earth's temperature would be -270° and no life would be possible. While the rays of the sun are indispensable to life on earth, they can also be dangerous for man and can destroy life. The sun is our best friend but it can also become our worst enemy.

Mercury

MERCURY is the smallest planet in the Solar System and the closest to the Sun. Mercury has a very thin atmosphere. Mercury is covered with craters from the bombing of asteroids.
The name Mercury comes from a Roman god renowned for his swiftness .

Position	1st place
Size	4,878 km diameter
Temperature	-180°C at night, 450°C by day
Satellite	No

Venus

Venus is the brightest of the planets after the Sun and the Moon. Venus is equivalent in size to Earth. For a long time scientists considered Venus and the Earth as twin planets. The atmosphere of Venus is essentially carbon dioxide that traps the heat of the sun. Venus is the warmest of the planets. It is this very high temperature that explains the brightness of Venus. Venus is so bright that it can be seen shining in the sky just before dawn or shortly after dusk.

Position	2nd place
Size	12 101 km diameter
Temperature	465C°
Satellite	No

Earth

Earth is theonly planet in the universe known to support life. It has an atmosphere rich in oxygen, and a large supply of liquid water on its surface. Earth sits at an ideal distance from the Sun if it were any closer, it would be so hot that water would boil away. If it were any further away, its surface would be cold and frozen
Almost three-quarters off Earth s surface is covered in Water.

Position	3rd place
Size	12 756.3 km diameter
Temperature	+58°C to -89,9°C
Satellite	the moon

The moon

The moon is Earth's only natural satellite that rotates around the Earth. it is even larger than Pluto and almost the same size as Mercury. The Moon is 384,000 km from the Earth, which has nothing to do with the size of the solar system.
The Moon is the nearest star to the Earth and the brightest after the Sun. The moon does not produce light, it only reflects the light of the sun. It is for this reason that it disappears completely when it is in the shadow of the earth. It is the eclipses of the Moon.

Mars

Mars is a cold planet and it freezes there almost permanently. Mars has always been known to men, the observation of sorts of canals on its surface led men to imagine that life existed on Mars. Mars is the planet that most closely resembles Earth. The landscape of Mars is very varied, there are mountains, valleys, canyons and craters. Mars has two ice caps composed of solidified carbon dioxide. It also has a fine atmosphere composed of carbon dioxide. Scientists believe there was water on Mars, but certainly for a very short period of time.

Position	4th place
Size	6 794 km diameter
Temperature	+22 c° to - 143 c°
Satellite	Phobos , Deimos

Jupiter

Jupiter is the largest planet in the Solar System. It is more than eleven times larger than Earth. In fact, this planet is so large that it could contain all the other planets in the solar system. Jupiter is by its size an easily observable planet of the Earth, it is therefore known since all time. Jupiter is a gas giant, made mostly of hydrogen and helium with a small rocky core Near the surface.
Jupiter was named after the king of the Roman gods

Position	4th place
Size	142 984 km diameter
Temperature	-145°C
Satellite	16 Satellite

Saturn

Saturn is the second planet by size. Saturn is primarily famous for its rings. The Voyager probe has revealed thousands of rings around Saturn. These rings are composed of millions of small fragments and particles ranging from the size of a grain to that of a large rock. Saturn's rings would be ice or ice-covered. The axis of rotation of saturn being inclined by 27 d° its visible form changes several times during its revolution around the Sun. Saturn is a sphere flattened on the poles. Of all the planets in the solar system Saturn is the flattest.

Position	5th place
Size	120 660 km diameter
Temperature	-160 c°
Satellite	23 Satellite

Uranus

Uranus is a large frozen planet composed of water. Uranus being a gaseous planet it has no solid surface. Uranus is the third largest planet in the solar system, 4 times larger than Earth. Uranus was not discovered until 1781 by the astronomer William Herschel. The discovery of Uranus is a coincidence: the astronomer was not looking for a planet, he thought he was simply observing a star among others. It is very hard to see from Earth with the naked eye, and it was the first planet to be discovered using a telescope. Uranus is named after the ancient Greek god of the sky

Position	6th place
Size	800 50 km diameter
Temperature	-200 C°
Satellite	27 Satellite

Neptune

Neptune is a planet of comparable size to Uranus, of dark blue color. This blue color is given by methane which absorbs the red light of the Sun. After the discovery of Uranus, scientists intrigued by certain anomalies deduced that another planet had to exist. Neptune was not discovered until 1846. Today all the knowledge we have of Neptune comes from Voyager 2. Neptune suffers many giant storms, winds can reach 2000 km/h on Neptune.

Position	7th place
Size	24 622 km diameter
Temperature	200°C
Satellite	14 Satellite

Pluto

PLUTO WAS DISCOVERED IN 1930, WHEN PLUTO BECAME THE LAST AND SMALLEST PLANET IN THE SOLAR SYSTEM. UNTIL AUGUST 2006 PLUTO WAS THE LAST KNOWN PLANET IN THE SOLAR SYSTEM. SINCE 2006 PLUTO NO LONGER HAS PLANET STATUS. FOR YEARS, SCIENTISTS WERE NOT PAV- ENUS TO CLASSIFY PLUTO IN ONE OF THE EXISTING CAT- EGORIES AND THEN, FINALLY, THEY DECIDED THAT PLU- TO DID NOT MEET ALL THE CRITERIA FOR CONSIDERED A PLANET. PLUTO WAS THE SMALLEST OF ALL THE PLANETS IN THE SOLAR SYSTEM. IT WAS ALSO THE STAR FARTHEST FROM THE SUN AND EARTH, IT WAS POORLY KNOWN. REDUCE

Position	9th place
size	2 300 km diameter
Temperature	-220 C°
satellite	Charon

Milky way

The galaxy which Earth belongs is called the Milky Way. The solar system is on the edge of the Milky Way. The size of our Milky Way is considerable and today we know only a very small part of our galaxy. The diameter of the Milky Way is estimated at 100,000 light years. The sun is only one of the hundreds of billions of stars of the Milky Way. The Milky Way is a spiral-shaped galaxy, so it is made up of arms that wrap around a group of stars placed in the center.

Comet

When it is visible, the comet appears as a star followed by a train. Comets are huge balls of ice and dust or rocks. The best description of the comets is that of the American astronomer Fred Whipple who describes them as "dirty snowballs". The comets formed at the same time as the solar system, so they are about 5 billion years old// The comets remain invisible as long as they are too far from the sun. Near the sun, the comet accelerates and becomes clearly visible by reflection of sunlight. Near the sun, the comet's ice melts and turns into a gas that forms the comet's head. The tail of the comet is formed by dust. At each passage before the sun, the comet loses a little

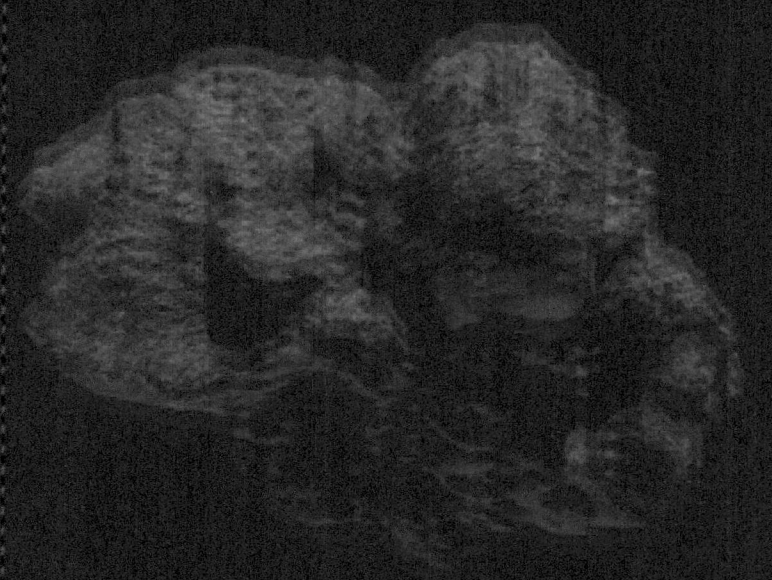

Asteroids

The asteroids are small rocky bodies gravitating around the sun, the vast majority of them gravitate around Mars and Jupiter. The first asteroids were discovered at the beginning of the 19th century. Today, scientists believe that asteroids are celestial bodies that have never clung together to form a planet. There is a great diversity of asteroids, by size, shape, color or composition. Asteroids are the subject of numerous researches because they were formed at the same time as the planets and their study will allow to better know the origins of the world

Copyright © 2020

All rights reserved. No part of this publication may be reproduced, distributed, or transmitted in any form or by any means, including photocopying, recording, or other electronic or mechanical methods, without the prior written permission of the publisher, except in the case of brief quotations embodied in critical reviews and certain other noncommercial uses permitted by copyright law.

www.ingramcontent.com/pod-product-compliance
Lightning Source LLC
Chambersburg PA
CBHW051935210526
45473CB00006B/2256